STEPHEN HAWKING

Stephen Hawking
His Science in a Nutshell

by Florian Freistetter
Translated by Brian Taylor

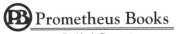

Prometheus Books

Guilford, Connecticut

PB Prometheus Books

An imprint of The Rowman & Littlefield Publishing Group, Inc.
4501 Forbes Blvd., Ste. 200
Lanham, MD 20706
www.rowman.com

Distributed by NATIONAL BOOK NETWORK

British Library Cataloguing in Publication Information available

Library of Congress Cataloging-in-Publication Data

Names: Freistetter, F. (Florian), author.
Title: Stephen Hawking : his science in a nutshell / by Florian Freistetter ;
    translated by Brian Taylor.
Other titles: Hawking in der Nussschale. English
Description: 1st US hardcover edition in English. | Amherst, New York :
    Prometheus Books, [2019] | Originally published in German: Hawking
    in der Nussschale : der Kosmos des grossen Physikers / Florian Freistetter
    (München : Carl Hanser Verlag, 2018). | Includes bibliographical references
    and index. | In English, translated from the German.
Identifiers: LCCN 2019013855 | ISBN 9781633885769 (hardcover) |
    ISBN 1633885763 (hardcover) | ISBN 9781633885776 (ebook) | ISBN
    1633885771 (ebook)
Subjects: LCSH: Hawking, Stephen, 1942-2018. | Physicists—Great Britain—
    Biography. | Cosmology.
Classification: LCC QC16.H33 F5813 2019 | DDC 530.092 [B] —dc23
LC record available at https://lccn.loc.gov/2019013855

∞™ The paper used in this publication meets the minimum requirements of
American National Standard for Information Sciences—Permanence of Paper
for Printed Library Materials, ANSI/NISO Z39.48-1992.

Contents

Prologue

WHEN I WAS SIXTEEN, I CHANCED UPON *A BRIEF HISTORY of Time* by Stephen Hawking. I hadn't been actively looking for the book, nor was I then particularly interested in natural sciences. But something about the book must have spontaneously caught my attention. I began reading the first chapter there and then in the bookshop, though I didn't yet have enough money to buy the book. It was only after a few more visits, during which I read another two chapters, that I actually took it home with me (after paying for it, of course) and read it through to the end in one sitting.

I have to confess that I only understood a fraction of the book's contents, something I probably have in common with most other readers. What I did immediately grasp, however, was that there was a fascinating universe out there, full of things that lay outside the scope of our normal understanding. Things like black holes, for example, out of which no information can escape and in which, nevertheless, are quite possibly concealed the answers to many of the Big Questions. The question about the beginning of the cosmos and its end, or the question about the nature of time. And the ultimate question itself: Why is there *something* and not *nothing*? The thing that impressed me most about the book, however, was something that seemed almost unfathomable at the time—the fact that *such questions could be investigated by scientists*.

It was only by reading Hawking's classic that I realized that the universe in its entirety is a research object for

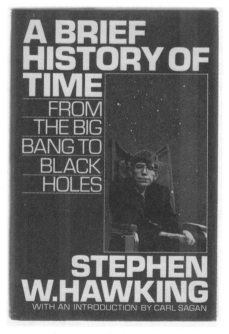

Cover of *A Brief History of Time* by Stephen Hawking

modern natural sciences and that there are physical theories and mathematical equations that are concerned with its formation, its characteristics, and its development. And even though I had no thought of understanding these theories and equations, I was gripped by the idea that it was possible to do so.

Reading *A Brief History of Time* was what made me decide to study astronomy. My lack of success in math at school made no difference—I knew that what my teacher taught me (in a rather poor and inadequate manner, as I now realize) was not what really mattered. In the math lessons, we only learned about calculating, which was boring. As a result, I didn't really make much of an effort. What I later learned during my astronomy studies at university, however, was not school mathematics—it was the language in which nature reveals itself to us; the language through which the mysteries of the universe can be understood; the language which Stephen Hawking used when he set out in search of the answers to all the big questions.

As it happened, I ended up not specializing in cosmology (the scientific study of the universe as a whole) and devoted myself instead to the motion of asteroids, planets, and other heavenly bodies. My scientific work was never anywhere near as intensely involved with black holes, the Big Bang, and the fundamental natural laws as Stephen Hawking's. But I did learn from him how fascinating the universe can be—and how wonderful it is to share this fascination with other people.

Few other scientists have succeeded like Stephen Hawking in making the beauty of the cosmos accessible to the wider public. For the whole of his scientific career, he didn't merely strive to find answers to humanity's big questions—above all, he also spoke about his work in such a way that as many people as possible could share in his findings. It makes little difference that the (mathematical) details of his research can scarcely be presented in an easy-to-follow manner and are barely comprehensible even after years of study. The physics genius with his futuristic wheelchair and penetrating, computerized voice knew better than anyone how to pass on the joy, fascination, and satisfaction that come from the study of the universe.

The discoveries I have made in my scientific career are not nearly as significant as those made by Stephen Hawking. But I too am completely convinced that it is important and, above all, extremely rewarding to share the findings of science with as wide an audience as possible. When I attempt in the following pages to make the most important results of Hawking's work understandable, therefore, I do so in the hope that they will continue to inspire as many people as possible to concern themselves with the big questions—and with natural sciences, the discipline that sets out in search of the answers to these questions.

I don't know how my life would have turned out if I hadn't chanced upon *A Brief History of Time* back then. But it is a source of great joy to me that Stephen Hawking's thoughts reached me at the right time and that I was able to get to know his cosmos.

Chapter One

Singularities

The Beginning of the Universe

Stephen Hawking started his scientific career with the ultimate beginning: the question of the genesis of the universe. Philosophers and theologians had concerned themselves with this matter for centuries, but in the twentieth century, the natural sciences also began to investigate the origin of the cosmos. It was above all Albert Einstein with his general theory of relativity who provided a tool that allowed the universe to be studied in its entirety, a tool that countless scientists would go on to use—including the young Stephen Hawking.

On October 18, 1966, the year when Hawking finished his doctoral studies at the University of Cambridge, he published an article titled "The Occurrence of Singularities in Cosmology," which was about the universe's past and the issue of "singularities." The latter term is closely linked to Einstein's space-time, one of the great thinker's many major achievements and something that

Young Stephen Hawking

still occupies scientists today. Before Einstein, people had kept to what Isaac Newton had had to say on the matter: space was space, and time was time.

The one was independent of the other—time was absolute and passed for all of us at the same rate. Time and space were the stage on which every single event in the universe was played out. Einstein, however, did away completely with this idea and demonstrated that the three dimensions of space and the single dimension of time are inextricably connected in the form of a four-dimensional space-time. Since Einstein, we have been aware that how space appears to us, and how we perceive time, depends on how quickly we are moving. In other words, time and space are not absolute terms but rather appear different to each observer. Einstein turned Newton's stage for the laws of nature into a physical entity: space-time is itself subject to physics—it has characteristics and can change. Above all, it can be shaped: mass and energy warp time and space, and we perceive the varying strength of this distortion as a difference in the strength of the gravitational pull.

All of that is confusing enough. When scientists took a more detailed look at Einstein's equations, however, the whole thing became even more complicated, for they came across singularities. They only began to understand these when they investigated the development of stars. These huge balls of hot gases have nuclear fusion taking place inside them. The energy thus released is emitted outward and pushes against the stars' matter.

This radiation pressure acts against the force of gravity, since the star actually keeps trying to collapse in on itself under its own weight. But when, at the end of its life, a star can no longer carry out nuclear fusion due to a

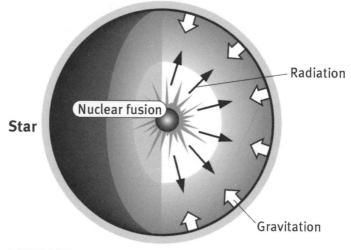

© PETER PALM

lack of matter, the pressure is removed. Now, gravity gains the upper hand, the star collapses in on itself and becomes ever smaller, with the matter of which it consists pressed more and more densely together. This collapse can be stopped when the atoms are closely packed and the star's gravitational pull is no longer sufficient to press them any closer together. If the star is sufficiently high-mass, however, there is no known force that can stop the collapse. The equations of the theory of relativity show that the star becomes ever smaller and denser, until its entire matter is finally united in a single, tiny point. The space-time around the dying star is increasingly warped during this collapse—until a point when this distortion and the star's density are *infinitely* great and the star itself is *infinitesimally* small. This state, in which physical dimensions become infinite, is called a "singularity."

If a star dies as just described, we refer to it as a "black hole," but we are not in a position to follow its development to when it ends up as a singularity. Why not? Well, it's because no more information about this end can reach us—when space-time becomes increasingly warped during the star's collapse, this results in an ever-increasing gravitational pull. The stronger an object's gravitational pull, the more energy is needed to distance oneself from the object, or the faster one must be. Take the Earth, for example—to escape its gravitational pull for good, you

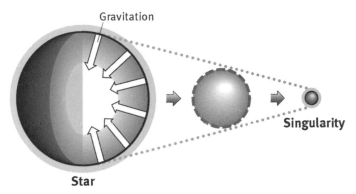

Gravitation

Singularity

Star

have to move at a speed of at least 11.2 kilometers per second. The greater the curvature of space, the greater this "escape velocity" is. With a collapsing star, a point will be reached when the velocity required equates to the speed of light, meaning it would be necessary to move quicker than light to escape the gravitational pull—and that's impossible. This limit is referred to as the "event horizon" and defines what we perceive from outside as a black hole. Up until the event horizon, it is possible to approach a black hole (and, if you're quick enough—that is to say, goddamn quick—to get away again), but beyond the event horizon you have no hope. This is why everything behind the event horizon is invisible to us outside. Nothing can escape from there, and therefore we do not know what is actually behind it. Relativity theory states that the star behind the event horizon continues to collapse, until it ends up as a singularity.

Nevertheless, the event horizon is in actual fact not nearly as mysterious as it sounds in theory. You cannot see it—it is not a real point in space, not a physical barrier. If you were to approach a true black hole, you wouldn't notice anything special when you crossed the event horizon. It's only when you wanted to leave again that you would have a problem. Or when you came across the singularity itself.

The scientists knew, of course, that singularities do not correspond to reality. Infinitely small objects cannot exist outside of mathematics. If singularities appear, this is a sign that the theory being used no longer works and so we need to come up with something else. In the 1960s, however, people still believed that we could, if necessary, simply ignore such cases. It was thought (and hoped) that the singularities arising from the theory of relativity were merely a kind of mathematical oddity, arising from certain assumptions made for simplification purposes when using the theory. This point can be explained using a less mysterious example: Coulomb's law is a mathematical formula that describes how powerful the electrostatic force of attraction is between two electrical charges. If the distance between the two charges equals zero, the formula states that the force of attraction is infinitely great. Here again, there is an infinitely great value, and here again, there is a singularity contained within the formula. But it arises from a mathematical idealization: this law describes particles such as atoms as points. Two points can indeed—mathematically speaking—get so close to each other that the distance between them equals zero. Real particles, however, are never points, that is to say, objects without dimensions. Instead, they are particles with dimensions that, while they may be very small, do actually exist. And the distance between real particles can never become zero. The singularity in Coulomb's law is indeed just the result of the implied assumptions made when using the theory. It was hoped that the singularities in the theory of relativity would be the same and that the death of stars could be explained without them.

But, returning to Hawking—the focus of his article about singularities wasn't on collapsing stars. He was interested in the much bigger picture, the entire universe itself, and he used collapsing stars to draw parallels with the development of the cosmos.

Observations of galaxies made by the American astronomer Edwin Hubble and his colleagues in the 1920s revealed the universe to be dynamic. Previously, people had thought that the cosmos was static—it had always been there and would always be there, with neither a beginning nor an end. But Hubble discovered that all galaxies were moving away from each other, that the universe was expanding and increasing in size, bit by bit, from one moment to the next. If we don't look to the future, however, but rather to the past, the situation is reversed. The further back we look, the smaller the cosmos is. But what happens if we go back really, really far?

Space-time becomes smaller and smaller, with more and more mass pressing together into an ever smaller space. In other words, the situation resembles that of a collapsing star. And just as we can use Einstein's equations to describe such localized areas of space-time, we can also calculate how space-time in its entirety behaves. This was the revolutionary step that Hawking was the first to take, and, here too, he came up against singularities.

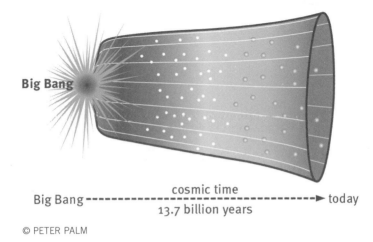

Big Bang

cosmic time

Big Bang --➤ today

13.7 billion years

© PETER PALM

If the universe was indeed smaller in the past than it is today, Hawking realized that there were two possibilities for its initial state. Either the universe was once in a state where its density had reached a (very large, but not infinite) maximum value. Or there was in the past a universal singularity—that is, a state in which the entire mass of the universe was concentrated in a single point: a point with infinite density, temperature, and space warp.

Scientists had discovered this unsettling situation in Einstein's equations back in the 1920s. Since the equations were incredibly complex, however, they could be solved only if very particular assumptions were made concerning the distribution of matter in the universe, the symmetry of the universe, and so on. It was now mooted—and hoped—that singularities would disappear from the theory if less particular, and therefore more realistic, assumptions about the characteristics of the cosmos were made.

This was where Stephen Hawking made his grand entrance in October 1966. In his article, however, he didn't demonstrate a way of eliminating singularities from the theory of relativity. Rather, he showed that it was impossible to escape them. He calculated that merely a few very sensible and realistic assumptions about the universe—that the general theory of relativity was correct, that the universe contained at least as much mass as could be observed—were sufficient to make us end up with the singularities again.

In other words, Stephen Hawking was able to show that the singularity at the beginning of the universe was not a mathematical oddity in the general theory of relativity. Gravity is a force that can only ever attract. This is one of the things stated by the general theory of relativity. If we assume that the theory is correct, as Hawking did in his work, then the singularities arise directly from this. They simply cannot be avoided in the general theory of relativity, no matter how much we try. They are inherent in the theory, and, if we work with it, then we have no choice but to come to terms with the singularities.

If we apply Einstein's theory to the universe, we end up with a situation in the past in which everything was concentrated in a single point of infinite density and temperature, a finding that is remarkable for more than one reason. First, it confirmed the "Big Bang" theory—the idea that the universe had a beginning some time in the past and had developed into the cosmos today from this original point. Second, it also provided even clearer evidence that the classic explanation of the universe's past did NOT always tally with Einstein's equations, since if you go far enough back into the past, you necessarily arrive at a singularity (with its infinite physical dimensions), with which his theory of relativity no longer worked.

Our universe is dominated by gravity, and because this is the case, as Hawking was able to demonstrate, we find singularities in the theory of relativity, in particular when we look into the past and wish to observe the beginning of the cosmos. Hawking showed both that the universe must have begun with a singularity and that we cannot rely solely on Albert Einstein. We have to find another approach for explaining the cosmos if we really want to understand how everything began. We need a theory that goes beyond Einstein's theory of relativity, a theory in which the Big Bang is no longer a singularity that is incomprehensible in nature and can instead be understood. And it was this search to which Hawking dedicated a large portion of the rest of his career.

CHAPTER TWO

Gravitational Waves

When Black Holes Collide

Stephen Hawking's work on the singularity at the beginning of the universe brought him fame as a scientist at the end of the 1960s and made a significant contribution to a clearer understanding of the cosmos. In the years that followed, however, he devoted himself to the objects that are most associated with his scientific work today: black holes.

He arrived at these via a phenomenon that is often overlooked in the multiplicity of the subjects he addressed. In 1970, he and Gary Gibbons published an article about gravitational waves, "Theory of the Detection of Short Bursts of Gravitational Radiation." Here again, we find Hawking operating in the field of difference between Isaac Newton and Albert Einstein. In his mathematical description of gravity, Newton had explained that it was a force that spread out at infinite speed. Einstein, whose theory described gravity as the effect of the distortion of space-time, begged to differ. He said that the distortion of space-time could not spread out infinitely quickly, but rather "only" at the speed of light. On top of this, he described *how exactly* changes in the distortion of space-time could spread out.

Gravitational waves can be compared to light waves. If light waves can be described as changes in an electromagnetic field that are spreading out, gravitational waves can be understood as changes in the distortion of space-time that are spreading out. However, they are much more difficult to substantiate than electromagnetic radiation. In popular science, gravity is often demonstrated with the help of a rubber sheet: balls of different weights, designed to represent the planets and stars, are placed upon a rubber sheet, which is then distorted to different extents, depending on the weight of the balls, just as real objects in the universe distort space-time. But the latter is in reality nowhere near as flexible as a rubber sheet.

Gravitational waves are produced by all (accelerated) moving masses, but it is only in the most extreme events that they are strong enough for us to have at least a theoretical chance of measuring them with our instruments.

Exploding stars or colliding black holes are needed for giant gravitational wave detectors like that of the American Laser Interferometer Gravitational-Wave Observatory (LIGO) to be activated. And this only actually happened in 2016, almost exactly one hundred years after Einstein first predicted the phenomenon.

You see, the quest to find gravitational waves was no easy task for scientists. When a gravitational wave moves through the universe, it stretches space itself. When it meets the earth on its way, the earth too becomes a little longer or shorter for a brief moment, though "longer" and "shorter" are relative terms here, since each measuring instrument is also similarly compressed. To prove the existence of this phenomenon, however, a technique called interferometry was used. Put simply, this involves a laser beam being created and then split with a mirror into two beams diverging from one another at right angles. After a certain distance, the two beams are each reflected in a mirror and travel back to the starting point. If the distances they have traveled are exactly the same length, the two beams arrive back at this point at exactly the same time, since they both move at the speed of light. If a detector has been installed at this point, it can be set up in such a way that the beams eliminate one another there.

If a gravitational wave hits the earth, however, the distances change. If the earth is stretched and compressed in different directions, the distance traveled by the laser beam will also be longer or shorter. But since there is not only one laser beam, but two, traveling at right angles toward each other, the change is not identical for both of them. If the one distance is longer, the other becomes shorter, and vice versa. In any case, a gravitational wave means that the two laser beams no longer arrive at the detector at the same time, cannot eliminate each other there, and instead create a signal that can be measured.

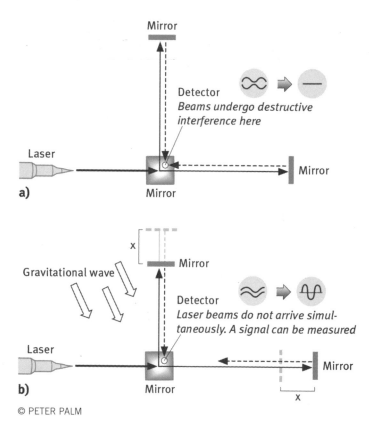

© PETER PALM

That being said, the effects of a gravitational wave are minimal. The LIGO laser beams both travel a distance of four kilometers before they are reflected. The change in the length of this distance, however, is thousands of times smaller than the diameter of an atom nucleus, and tremendously sophisticated technology is required to detect such an infinitesimal effect. Above all, we have to be certain that we have truly detected a gravitational wave and not something else, like a car passing by, an earthquake somewhere in the world, or some other source of disturbance. It is for this reason that LIGO consists not only of one facility but rather of two identical detectors located three thousand kilometers apart. It is only when both of these detect the same signal that we know we are dealing with an astronomical phenomenon.

The planning, construction, and testing of LIGO took decades, but all of this work finally proved its worth on September 14, 2015. That day, the first gravitational waves were detected, and, after the findings had been comprehensively evaluated, the discovery was officially announced on February 11, 2016.

As early as 1958, the physicist Joseph Webber had announced the detection of gravitational waves—using a much simpler experiment setup. Even today, it is still not completely clear whether what he thought he had detected at the time was down to a measurement fault or whether he had indeed discovered gravitational waves (with majority opinion tending toward the measurement fault idea). Nevertheless, theoretical physicists investigated Webber's measurements, and Hawking himself speculated in the 1970 article already mentioned about detection methods that would clear the matter up. He calculated how strong the gravitational waves that arose during astronomical phenomena might be and suggested how suitable detectors might be constructed. He also suggested doing what LIGO then did—building more than one detector in order to be able to find out the exact source of the signal.

A year later, another article by Hawking was published, "Gravitational Radiation from Colliding Black Holes." It too was concerned with gravitational waves, but it went far beyond the issue of their detection. In this article, Hawking published what is known today as his "area theorem."

When two black holes collide, like the event that led to the first evidence of gravitational waves in 2016, a new black hole is created with a greater mass. The greater the mass, the greater the event horizon. And Hawking was able to prove that the event horizon of the black hole created by the fusion is always greater than the sum of the areas of the event horizons of the two individual black holes.

He established that black holes can indeed emit gravitational waves when they collide. They thus lose energy, and the single black hole created by the collision has less mass than the two original black holes together. The entire surface area that makes up the event horizon, however, cannot shrink. Hawking's mathematical derivation made it abundantly clear that, whatever physical processes may occur, the event horizon of a black hole never becomes smaller, something which recalls a completely different physical discipline, namely thermodynamics.

This contains four fundamental statements, the so-called laws of thermodynamics. The most important of these is the second one, which deals with entropy. Simply put, the entropy of a system can be defined as a measure of its disorder. A book, for example, is a very orderly system; each page has precisely one correct place, and there is precisely one arrangement in which all the pages are correctly numbered one after the other. On the other hand, there are many, many more ways of wrongly ordering the pages of a book. If you were to throw a pile of unsorted but numbered pages into the air, it is extremely unlikely that they would land on the ground in the single correct order. That would correspond to a change in the system, by which it would change from a state of high entropy (disordered pages) to one of low energy (ordered pages). It is much more probable that you would instead get one of the many other disordered states. If you wanted to turn the disordered state into an ordered one, you would have to invest energy in the system that is the book and sort the pages by hand. Otherwise, you would have no chance of reducing the entropy. The second law of thermodynamics states that the entropy of every isolated physical system can never decrease. That being said, completely isolated systems exist almost exclusively only in theory or for short periods of time. Hot coffee in a thermos flask, for instance, comes close to being an isolated system for some time, with no energy being transferred between the inside and the outside of the flask. In the end, though, a transfer does take place, and the coffee will eventually become cold. The only truly isolated system in physics is the universe in its entirety.

Entropy is similar to the surface area of a black hole's event horizon. Both of them can never become smaller by themselves. What leads to problems, however, is the fact that black holes themselves must have entropy, otherwise, they would contravene the second law of thermodynamics. To get our heads around the problem thus arising, we can carry out a thought experiment: sticking to the simplified idea of entropy, that of disorder, let's imagine a *really* untidy room. Yes, we could clear it up, but we would use energy in doing so and would have to make an effort and transmit the energy used into the surrounding environment in the form of heat, which would increase the disorder of the particles in the air and thus counteract what little order we had brought about by clearing up. Or we could spare ourselves the clearing up and simply throw the whole room into a black hole. Then the whole disorder, and with it the entropy, would be concealed behind the event horizon and inaccessible for the rest of the universe. While we would now have lost our room, we would effectively have reduced the universe's entropy, and this would contradict the second law of thermodynamics. Complicated stuff!

In 1972, the theoretical physicist Jakob Bekenstein discovered a creative way around this dilemma. He took the (until then only formal) correspondence between the event horizon of a black hole and entropy literally and *defined* the entropy of a black hole as simply the surface area of the event horizon (multiplied with a few fundamental constants). This was an unexpected move, since the fact that two formulae are similar to one another does not necessarily mean that they have an actual physical connection. But that was precisely Bekenstein's suggestion—if you throw something into a black hole, its mass, and thereby its event horizon, becomes bigger. If you define the surface area of the event horizon as its entropy, then the latter also becomes bigger, indeed at least so much bigger that the entropy that has disappeared behind the event horizon is balanced out. Our untidy room may have disappeared into the black hole forever, but the disorder itself is retained by the universe, in the form of an enlarged event horizon.

Initially, Hawking was not a fan of Bekenstein's idea at all but was then won over by its usefulness, going on to take the analogy between the characteristics of black holes and thermodynamics still further. Together with John Bardeen and Brandon Carter, he published an article in 1973 called "The Four Laws of Black Hole Mechanics," in which they formulated four statements about black holes that can be considered to be analogous to the four laws of thermodynamics.

Besides the statement about entropy, another of the laws of thermodynamics describes the retention of energy—the internal energy of an isolated system must remain constant. Hawking and his colleagues described a similar connection for the internal energy of a black hole, which can also not simply change at will. A further law of thermodynamics states that a physical system can never be cooled to absolute zero; the corresponding statement about black holes describes a similar characteristic for the strength of the gravitational acceleration at the event horizon, which can also never entirely disappear. And just as temperature differences of a physical system are balanced out over time, this is also true of the gravitational acceleration at the event horizon of a black hole. If we allow all external disturbances to subside, the rate of acceleration must be the same everywhere.

This relationship between black holes and thermodynamics is fascinating—but also confusing. Is it actually just a formal correspondence between mathematical formulae that happens to work very well? Or are we actually seeing here fundamental connections between two phenomena that at first sight appear to have nothing to do with one another?

Above all, there was one problem: Bekenstein's equating of the surface area of the event horizon with entropy had, along with Hawking's area theorem, demonstrated that black holes did not contravene the laws of thermodynamics. But if black holes were *really* objects that were subject to thermodynamics, then they must too have a temperature (since nothing can be cooled to absolute zero). And if they have a temperature, then they must give off heat, or radiation. Yet that is precisely what black holes, by definition, do NOT do!

Quite a dilemma—but one that Hawking was able to free physics from with a discovery that remains today one of his most significant contributions to theoretical physics.

CHAPTER THREE

Hawking Radiation

Why Black Holes Aren't as Black as We Thought

IF BLACK HOLES ARE NOT TO CONTRAVENE THE LAWS OF thermodynamics, they have to emit radiation. But they are called *black* holes precisely because nothing can escape the force of attraction behind the event horizon. It's a paradox—a seemingly insoluble problem. But Stephen Hawking, in a 1975 article called "Particle Creation by Black Holes," made a truly astonishing discovery.

He was able to demonstrate that black holes are not as black as was previously thought. His earlier work had paved the way for this discovery: when he proved that singularities are an unavoidable consequence of the general theory of relativity, it became clear that the theory was incomplete. If we wanted to explain extreme states of space-time like the Big Bang or a black hole, Einstein's ideas alone could only take us so far. But perhaps taking into account other theories would make it possible to solve the problem of a black hole's temperature, and that is precisely what Hawking attempted to do by turning to quantum mechanics.

We can see here, however, a further problem associated with this kind of research. It is almost impossible to find clear explanations. Hawking's work is purely mathematical, and its full scope can only be understood if you understand the mathematics he uses. It is frequently possible to find a descriptive approximation of mathematical formulae. But equally this is often not the case—especially when the mathematics is explaining phenomena that play no role in our everyday lives. We human beings have nothing to do with extremely warped space-time in our day-to-day lives, nor do we have an intuitive sense for how an event horizon functions or elementary particles behave. That is precisely why we have developed the mathematical definitions of the natural sciences—in order to be able to consider, in an objective and comprehensible way, things that we couldn't think about without mathematics.

Understanding mathematics to such an extent that you can follow the phenomena it explains is just as difficult as learning to speak a foreign language with perfect fluency. Not all of us can spare the time for this—nevertheless, we still want to know what scientists like Stephen Hawking have found out about the universe. Translating the findings of theoretical physics into a form that can be grasped without a profound knowledge of mathematics is difficult. It is necessary to simplify things and find analogies, and this necessarily means that a part of the actual information is lost.

Stephen Hawking found quite a vivid way of describing the radiation of black holes in *A Brief History of Time*. This representation has since been constantly disseminated, and simplified still further, so that we end up with a description that we can easily understand, but which has little to do with the actual phenomenon in question. The usual explanation of the "Hawking radiation" of black holes used by popular science today goes something like this: The empty space between the stars is not completely empty. So-called virtual particles can arise here. That may sound strange but is actually quite normal and happens everywhere in the universe. Pairs of particles are constantly popping up—one made of matter and one made of antimatter. The two particles exist only for a very brief time, so brief in fact that they don't actually exist and they then destroy each other. These virtual particles cannot be directly observed, but the effects of their existence have been proved in experiments. When these particles happen to stick their (metaphorical) heads on the event horizon of a black hole from space, one of them can cross this frontier and doesn't come back. The partner particle on the other side of the event horizon now has nobody with which it can be mutually destroyed, so it doesn't disappear and instead floats about in space, while the other one has to remain forever in the black hole. The particle that has fallen into the black hole has a negative energy, which reduces the mass of the black hole. From outside, therefore, we can see how the black hole becomes a bit lighter and, at the same time, how a new particle flies out into the world from the event horizon. The result? The black hole is "radiating," and everything is in keeping with thermodynamics.

This explanation is quite similar to the one Hawking himself used. But not completely the same. It has the advantage that we can easily imagine how Hawking radiation works, which is why people (including scientists) like to use it. It also has the big disadvantage that, as a description of the actual phenomenon in question, it is inadequate or even wrong. Take the bit about the negative energy, for instance—why does the particle that falls into the black hole have negative energy? What is "negative energy" even supposed to mean? Why doesn't a normal particle with positive energy sometimes fall into the black hole and thus balance out the negative energy? If we don't understand what is meant here, the whole phenomenon remains incomprehensible. Which is probably why many popular science books also say that it must always be the anti-particle that falls into the black hole and that's why the hole's mass is reduced (which is in fact a completely erroneous explanation).

The explanation from Stephen Hawking's own pen is somewhat more precise—though it is no longer as simple and, to be honest, is scarcely comprehensible without further illustration. When it comes to explaining Hawking radiation with pairs of particles at the event horizon, for instance, it would appear that this radiation always arises directly at the event horizon. This is not the case, however. The mathematical observation, which is the only valid one with this phenomenon, shows that Hawking radiation can also arise at a certain distance from the event horizon. The black hole is surrounded, as it were, by an "atmosphere" of Hawking radiation, and this phenomenon can no longer be explained by the pairs of particles at the event horizon.

For this reason, it would perhaps be useful to find a completely new image to explain Hawking's discovery. To do this, we first need to clear up two questions: What is a particle? And what is space? And, as we will see, the existence of Hawking radiation follows from the fact that there are no definitive answers to these questions.

Let's start with the first question. In quantum mechanics, particles are no longer described as the "little balls" that we still usually imagine them to be but rather as the excitation of fields. You can imagine the universe to be crisscrossed by fields, like the electromagnetic field. If you put enough energy into such a field, this excitation can produce a "particle." In the case of the electromagnetic field, this would be a photon, a light particle. But there are also "matter fields," such as an electron field, which would produce an electron when agitated.

However, because what we think of in quantum mechanics as "particles" are not particles and are actually fields, it is impossible to assign an exact position to the said particles. Put very simply, since the field is everywhere, each particle is also everywhere to a tiny extent. Only when we calculate a concrete measurement of the particle's position do we know where it is. But if we haven't exactly established this position, the measurement leads to our not being able to say how fast the particle is moving. This is the famous Heisenberg's uncertainty principle, which states that the more precisely the position of a particle is determined, the less precisely its momentum can be known, and vice versa. This uncertainty also applies in other cases. For instance, it isn't possible to simultaneously determine how much energy there is in a field and how quickly it is changing. Or to put it another way—a field can never simultaneously be completely gone (which would correspond to an energy of zero) and not change over time (which would correspond to a rate of change of zero), since we would then know both values exactly. Consequently, there must always be a bit of "unsettledness" with every field. Every field is constantly fluctuating a little, sometimes with more energy, sometimes with less. When the energy in these fluctuations happens to be great enough, it can be sufficient to produce particles (since, according to Einstein's famous formula $E = mc^2$, energy and mass are practically the same thing).

Particles that thus arise in space are referred to as virtual particles and always arise in pairs—a particle and an anti-particle, which eliminate one another almost immediately after spontaneously coming into being.

Which brings us to the second question. Because particles in quantum mechanics are described via the excitation of fields, space isn't simply "nothing" but is rather full of fields, containing more or less energy and out of which sometimes more and sometimes fewer virtual particles arise.

Incidentally, just because these particles are referred to as "virtual," this doesn't mean they are purely a product of the imagination. We know that they are there, because

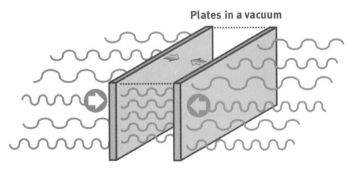

Plates in a vacuum

© PETER PALM

their existence can bring about effects that can be observed, such as the Casimir effect, a phenomenon predicted by the Dutch physicist Hendrik Casimir in 1948. Let's imagine two plates in space. Normally, we would expect the plates simply to remain where they are. If space were indeed "nothing," then there would be nothing there which might cause them to move. But quantum mechanics states that virtual particles are constantly arising everywhere in space. And since quantum mechanics also states that particles are actually waves, this means that it isn't possible for just any old particles to arise in the space between the plates—it can only be those whose wavelength fits exactly between the two plates. Outside the plates, on the other hand, there is enough space for all possible kinds of particles. In other words, fewer virtual particles can arise between the plates than outside them. As a result, more particles "press" against the plates from outside than from inside—and this leads to a force that makes the plates move toward each other. This effect was indeed observed in an experiment for the first time in 1956, and several other quantum mechanical experiments have since confirmed the existence of virtual particles.

Quantum mechanics tells us, therefore, that space is not simply "empty." Rather, it is full of quantum fields, out of which particles arise. So now let's take a look at the theory of relativity. Albert Einstein didn't just demonstrate that space and time are linked to one another—he also showed that we cannot ignore the state of motion of the observer. Depending on how quickly and with how much acceleration we move, time passes at a different rate. And this has an effect on how we view space. The more energy there is in the fields in space, the more particles arise there. The amount of energy, however, depends on how we observe time, since it is the energy-time uncertainty principle described earlier that is responsible for the formation of the particles from space. Since we can never simultaneously know exactly both how much energy there is in a field and to what extent a field is changing over the course of a particular period of time, the quantum fields must always be fluctuating.

According to Albert Einstein, observers who are moving in relation to one another at different rates of acceleration have different perceptions of time. It follows from this, however, that different observers are not in agreement as to how much energy there is in space. What one observes depends on how quickly one is moving. What is true for highly accelerated motion is also true for highly warped space-time, this too being one of the major findings that result from the general theory of relativity. If we move through warped space-time, we are subject to gravitational forces, the exact same forces that we feel when we are pressed back against the seat of an accelerating car. Gravity is nothing more than the way in which we perceive the warping of space-time. And acceleration and gravity are equivalent to one another.

When it comes to everyday forces and accelerations, the curious effects are limited, but near a black hole, things are different, since there, space-time is warped to an extreme degree.

To understand what all that has to do with Hawking radiation, we have to realize that a black hole is not a static object. We like to think of it as a thing that is either there or not. But this is wrong. A black hole is a dynamic process which describes space-time that is collapsing in on itself. If at first there is no black hole and then—perhaps because a star reaches the end of its life—one is formed, the result is not only extremely warped space-time, but also, and above all, a difference between past and future. Through this dynamic space-time, two observers move—one in the past and the other in the future—no longer in unison, but rather accelerated in relation to one another.

This leads, however, to a different view of space, energy, and particles. The observer in the past, before the black hole has started to interfere with space-time, sees just "normal" space with little energy. The observer in the future, however, sees space with lots of energy-filled particles. Hawking radiation, therefore, is essentially that into which the black hole has turned the space which was there before it was formed.

The virtual particles that are formed near warped space-time on account of the theory of relativity and the quantum fluctuations in space are responsible for black holes emitting radiation. The more strongly the space-time is warped, the greater this effect is. The warping is, however, slighter with a black hole with a big mass and thus a correspondingly large event horizon than with a small black hole. We can illustrate this using the earth: here too the surface is curved, but since our planet is so large in comparison to us, we barely notice the distortion and the ground appears flat to us. The curvature of the surface of a much smaller football, on the other hand, is easily perceived. Black holes with a large mass (and a large event horizon) create less Hawking radiation than small ones.

In the end, black holes are thus revealed to be fully consistent with thermodynamics. They emit radiation with a temperature that depends on their mass, and even if their event horizon shrinks while doing so, the entropy in the radiation ensures that the second law of thermodynamics is not contravened.

Even this more detailed (and less easily understood) explanation is still just an approximation of the complete mathematical description. But the phenomenon itself is no longer disputed in science today. Thanks to Hawking radiation, black holes are not as black as was previously thought. They do shine, at least to a tiny extent, though the word "shine" is actually a massive exaggeration in this case. The radiation emitted by a typical black hole is so incredibly low that it cannot be detected by any measuring instrument that we can devise.

The radiation does ensure that a black hole continually loses mass over time, until it has completely evaporated (at least we assume so, since we still don't know what really happens when a black hole nears its end—see chapter 4). But until this happens with a typical black hole, an unimaginable length of time must pass, since it not only emits Hawking radiation but also continually takes on matter or energy. There are always a few atoms somewhere in interstellar space that come across an event horizon on their way through the universe and are swallowed by it. And even if this doesn't happen, there is always the cosmic background radiation.

The latter can be viewed as the first light of the cosmos. When the universe began some 13.8 billion years ago in the singularity of the Big Bang (or when something happened that we can only currently describe as a "singularity" thanks to the inadequacy of the general theory of relativity, as Hawking demonstrated), the cosmos was still too hot and dense for anything like normal matter to be able to exist. In the universe's initial stages, there weren't even individual atoms. These normally consist of a nucleus surrounded by a cloud of electrons, but the temperature was still so high back then that the electrons moved too quickly to be permanently attached to the nuclei. They rushed freely through the universe, and their presence prevented the light particles from spreading out—these were constantly being diverted by the electrons, so that the cosmos was an opaque brew of light particles and incomplete atoms. It was only when, after approximately 380,000 years, everything had cooled sufficiently for the electrons to bind with the atomic nuclei that the way became clear for the light. Since then, this first radiation has been spreading out in the universe, and part of it is still going today.

When the physicists Arno Penzias and Robert Wilson proved the existence of this radiation in 1965, it was one of the strongest pieces of evidence for the Big Bang theory. Today, the radiation has cooled dramatically and has a temperature of only -270 degrees Celsius, but it is still to be found everywhere in the universe, and some of it is swallowed by the black holes. The small amount of energy obtained from the background radiation is easily enough to balance out what the black holes lose via Hawking radiation.

For the mass of a black hole to actually be diminished by the production of Hawking radiation, the cosmic background radiation first needs to become significantly weaker than it currently is. This will take a long time, however. The further the universe spreads out, the cooler the background radiation becomes, and only in the distant future will it be weak enough to allow black holes to begin to dissolve. It takes about 10^{68} years (i.e., one hundred undecillion years, so 1 followed by 68 noughts) for a typical black hole to disappear. That is such a long time that we cannot even conceive of it—so unimaginably long that, in comparison, the universe's lifetime so far of 13.8 billion years can be described as a mere blink of an eye. There is, therefore, no chance of directly observing the dissolution of a black hole and thus the existence of Hawking radiation—or at least the radiation that is emitted by black holes that are formed when stars collapse.

Theoretically, you see, there could also be smaller black holes which could have been formed when there was nothing more than a "stew" of energy and elementary particles in the very early universe. Bits of the universe which had an increased mass or energy density on account of the quantum mechanical fluctuations could have collapsed to form a black hole back then.

These microscopically small black holes might have a tiny mass equivalent to that of an elementary particle. They might also be heavier, but perhaps weigh "only" as much as a small mountain. Such black holes would dissolve more quickly than those formed from stars and would give off more Hawking radiation. *If* they existed, they could release large amounts of high-energy radiation when they finally dissolve, and should this occur anywhere in our Milky Way, then we should theoretically be able to observe it with suitable telescopes. So far, however, this has not yet happened.

Black holes that are even smaller again could, under the right conditions, be produced in experiments with particle accelerators, where particles are made to collide with great energy. It must be stressed that these mini black holes would be completely harmless, since they would dissolve in a fraction of a second on account of their small mass and the strength of the Hawking radiation. Their dissolution could however be observed by the particle accelerator's detectors. But this too has not yet happened, and it would appear that much larger accelerators would be needed in order to obtain such observations. Indeed, the lack of experimental proof for Hawking radiation is probably the reason why Stephen Hawking was never awarded the Nobel Prize for his work.

Nevertheless, he did demonstrate the incredible findings that can be obtained when we combine quantum mechanics with relativity theory. In his first article on singularities, he showed that we can only go so far with relativity theory alone, with difficulties encountered in situations like the Big Bang or the explanation of black holes. By using quantum mechanics, he himself was able to solve some of these problems. Quantum mechanics, along with relativity theory, the second major theory of modern physics, is extremely successful when it comes to explaining the world of atoms and elementary particles. But it is a theory without gravity; this fundamental force of nature does not appear in it. Hawking combined quantum mechanical effects with the consequences of the general theory of relativity and thus came up with the radiation that bears his name. His theoretical explanation was, however, a long way from being a complete amalgamation of quantum mechanics and relativity theory, and that is precisely what is needed if we wish to understand what really happens in black holes or what we can really imagine the Big Bang to have been. And there is still plenty to find out when it comes to these two questions.

Chapter Four

The Information Paradox

It All Continues behind the Event Horizon

By discovering Hawking radiation, Stephen Hawking managed to solve a major problem and made a significant contribution to our understanding of black holes. Nevertheless, these fascinating objects have still not by any means yielded all of their secrets—and one question that is still to be answered is based precisely on the fact that, at first glance, there doesn't seem to be much that needs to be understood about black holes.

We're talking about the "no-hair theorem," which states, succinctly and unsurprisingly: black holes have no hair. The theorem originates from the American physicist John Wheeler, whose aim was to show how uncomplicated black holes are. You see, hair has to be washed, combed, and cut, and there are all manner of possible styles. By choosing our hairdo, we humans can set ourselves apart from other people (assuming we still have enough hair to do so). Black holes, on the other hand, can't do this. It makes no difference what material they were formed from, or how complex the matter that has disappeared is. If you throw an eighty-kilogram man into a black hole, he disappears, and the mass of the black hole has increased by eighty kilograms. If you chuck an eighty-kilogram bag of cement into the hole, the result is exactly the same. There is a great difference between a

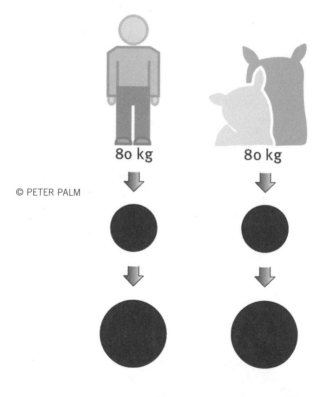

80 kg   80 kg

man and a bag of cement, but once they have both disappeared behind the event horizon of a black hole, this difference is evened out.

The investigations into the collapse of black holes carried out by Stephen Hawking and Roger Penrose clearly show that it makes absolutely no difference what characteristics are possessed by matter that collapses to form a black hole. Once it *has* collapsed and the event horizon has been formed (and it is stationary—that is, at rest, with all other influences and oscillations having subsided), then all we can still perceive from outside is its mass, its electrical charge, and its angular momentum. A black hole has precisely these three characteristics—more cannot be known about it. It has "no hair"—there is no way of further "individualizing" separate black holes. Hawking and Penrose seem thereby to furnish evidence for the "no-hair theorem," although the term "no-hair assumption" would actually be more appropriate, since the theorem has been mathematically proven only under very particular circumstances. The scientific community is not in agreement as to whether black holes in general are indeed as indistinguishable as they appear. In some hypotheses concerning the unification of quantum mechanics and the theory of relativity, the theorem no longer works, for example, where a universe with more than just three spatial dimensions is assumed. Since nobody actually knows whether these hypotheses are correct, however, little can be said about the general validity of the no-hair theorem. But if black holes did indeed have no hair, as it were, then we would once again have a problem, since information could then be destroyed in a black hole—and such a thing cannot happen.

Let's imagine that I were to burn a book. In practice, I end up with a pile of ash and a load of smoke, and all the information contained in the book is lost. This is correct in practical terms but not quite accurate in theory, since I could theoretically record how each individual ash and smoke particle has moved, reverse the process, and thus create the original book out of the ash. Naturally, that would be far too complicated to carry out in practice, but it would work in theory, since all known laws of physics can, at least in principle, be reversed. They do not favor a particular direction as far as time is concerned. By burning the book, I would have made a major change to the information therein, but I would not have destroyed it.

This principle is also true in quantum mechanics. As shown in the previous chapter, quantum mechanics is full of "uncertainties," but it is nevertheless *deterministic*. This means that, if we have a wave function (the abstract mathematical object with which quantum mechanics describes particles), then the future state of the wave function is determined by its present state. And the mathematical procedure with which we calculate the future development of the wave function can also be reversed. So here too information cannot be destroyed.

But what if I were to throw a book into a black hole? It would disappear behind the event horizon and never return. This development can no longer be reversed. A black hole thus seems able to destroy information, and this contradicts the fundamental laws of known physics.

We could console ourselves with the thought that, while the book may now be out of our reach, it is at least still there behind the event horizon, and with it the information. But the event horizon of a black hole is not simply a "curtain" that conceals what is behind it. It exists because there is behind it an extremely distorted area of space-time (if this were not the case, there would be no event horizon). The matter of which the black hole consists must have been compressed to such an enormous degree that it is essentially now just an unimaginably dense "point" of matter. In this extreme space-time, the book must necessarily be split into its component parts and become a part of this extreme state of matter.

From outside, we can observe only how heavy the black hole is, what electrical charge it possesses, and how great its angular momentum is. With these three characteristics alone, we haven't got a chance of finding out what matter is to be found behind the event horizon. The question that now arises is: can we perhaps after all reconstruct the information that has disappeared into a black hole?

Stephen Hawking was able to demonstrate that even black holes emit radiation. They also dissolve, even if only extremely slowly. So when the black hole has evaporated after an unimaginably long time, does the book appear again? No. First of all, it was destroyed. And second, the matter doesn't simply hide behind the event horizon and wait to see what happens. If it emits Hawking radiation, then this makes the black hole's mass smaller and smaller. Whether this mass completely disappears in the end, or whether something is left behind after all, is not yet known, but the book could in any case no longer be reconstructed.

That being said, the Hawking radiation could here play the role taken on by smoke and ash when the book is burned. Careful analysis of the burned remains would, at least theoretically, permit a reconstruction of the book and the information contained within it. If the Hawking radiation could be precisely observed, might we find variations therein that would depend on, and point to, the kind of matter that had disappeared into the black hole?

Stephen Hawking spent a long time investigating this question, and his initial answer was "no." If the no-hair theorem is correct, then it doesn't matter *how* the matter out of which the black hole is formed was originally created. In the end, we are left with a thing that has precisely three characteristics and not more. That's why Hawking radiation must also be purely "thermal," that is to say, completely independent of the matter that falls behind the event horizon. Even if it were technically possible to observe the Hawking radiation of black holes, we wouldn't find any information that was in any way connected with the lost book. That's what the early Hawking thought.

To begin with, therefore, he was convinced that black holes could indeed destroy information. Other physicists saw things differently. This was hardly surprising, since there wasn't at the time (and still isn't today) a theory that united quantum mechanics and relativity theory and thus enabled a comprehensive understanding of black holes. All we had and have is a load of hypotheses about what such a comprehensive theory might be like, and, depending which hypothesis we support, we can use it to solve the information paradox or not.

One of the preferred solutions for this information paradox bears the wonderful name "the anti-de Sitter/conformal field theory correspondence," or AdS/CFT correspondence for short. This linguistic monstrosity is great if you want to show off at a party (or scare people off, depending on your audience). A detailed explanation of what it is actually about would go far beyond the remit of this little book, however, so to put it very simply, it refers to a "duality"—that is, two different theories that both explain the same phenomenon, though from different perspectives. You can compare it to the use of two tools—if you can't solve a problem with one of the tools, perhaps you'll have more luck with the other. On the one hand, the AdS/CFT duality concerns a *quantum field theory*, the kind of theory that we currently use to explain elementary particles and the forces between them. The second tool in this duality consists of several theories at once, all originating from hypotheses for the unification of quantum mechanics and relativity theory. There seems to be similarities between these two ways of explaining the world, and if the two approaches were indeed to conform to one another, then we could derive from them statements about the possible loss of information in black holes. We could then deduce from this duality, for instance, that Hawking radiation is not completely uniform after all but actually exhibits minimal "quantum fluctuations" with which we could reconstruct information about the matter lost behind the event horizon.

Hawking himself also spent time working on this AdS/CFT correspondence and changed his opinion about the information paradox. In July 2004, he came to the conclusion that information in black holes does NOT get completely destroyed, in the process losing a bet that he had previously made on the subject. In 1997, he and his colleague Kip Thorne had made a wager against the physicist John Preskill that solving the information paradox could be done only by correcting quantum mechanics, since Hawking radiation itself could not transmit any information from inside a black hole. Preskill, on the other hand, was convinced that quantum mechanics allowed for the transmission of information through Hawking radiation and that the paradox must instead be solved via a change in the general theory of relativity. The findings from the AdS/CFT correspondence caused Hawking to concede that Preskill was right, with the latter becoming the proud owner of a baseball encyclopedia for winning the bet. Hawking later joked: "I gave John an encyclopedia of baseball, but maybe I should have just given him the ashes."

Despite all this, the subject is far from closed. The information paradox is one of the most fascinating unresolved issues that Stephen Hawking left to the world. There is still no theory that tells us definitively how black holes work and what happens to information that disappears behind an event horizon, though there is no shortage of hypotheses, as noted before. Perhaps information is destroyed after all. It may contradict our human sense of aesthetics to accept that information can simply disappear, but the universe has no obligation to please us. But perhaps something is left over when a black hole has dissolved, some strange quantum object that we cannot imagine with the theories of today but which has saved all the information about the matter that has fallen into the hole. Or perhaps the information simply ends up in a parallel universe. We will truly understand the information paradox only when we have a theory that tells us what the deal is with the singularity that is to be found behind the event horizon. So this legacy of Hawking's is sure to keep us busy for a long time yet . . .

Chapter Five

Before the Big Bang

In the Infinite Expanses of Euclidean Space-Time

In 1983, after the success brought on by the connection of relativity theory and quantum mechanics with regard to black holes, Stephen Hawking turned this approach to the major singularity that he had come across at the beginning of his career. Nobody had any understanding of the moment of the Big Bang. Existing scientific theories provide a very good description of the development of the universe, and we can use them to look far back into the past—13.8 billion years back, almost to the moment when the singularity, the Big Bang, appeared. Shortly after that, the cosmos became explainable for us, and we can work out how the universe developed from its initial state. The predictions we gain from these theories conform nicely with what we can actually observe. When we observe the background radiation mentioned in chapter 3, it looks just like the theory predicts. The amounts of chemical elements in the universe are also exactly what they should be according to the calculations. Naturally, though, there are still plenty of unanswered questions about the development of the universe.

Take the existence of matter, for instance. Why is there so much more matter in the universe today than antimatter? According to the theoretical models, matter and antimatter should have arisen in the same amounts at the beginning of the cosmos and should actually have then eliminated each other. The universe would then have been full of energy, but devoid of matter. Obviously, however, there must be some difference between matter and antimatter and a process that gives preference to one over the other, but we have yet to understand this. The question of cosmic inflation is also still unanswered. The Big Bang theory stipulates that the entire cosmos, very soon after its formation, expanded to an extremely great degree within an extremely short period of time—unimaginably more quickly than ever before or after. Even today, though there are many clues that suggest this was the case, there are still no concrete and unequivocal observations that provide evidence of this phase of the early universe (though there could soon be, since gravitational waves must also have been produced back then and we may be able to substantiate these with future detectors). Another major mystery concerns the nature of "dark matter." For almost one hundred years, we have been able to observe how stars and galaxies in the universe move in a fashion that cannot be explained only by the gravitational pull of the visible matter. There must be an additional source of gravitational pull, probably a still-unknown kind of matter that has not yet been identified or directly verified in experiments. "Dark energy," while also "dark," has nothing to do with "dark matter." To their great astonishment, astronomers in 1998 observed that the rate at which the entire universe is expanding is constantly increasing. There has to be a reason for this, even if it is still completely unknown. There is therefore plenty left for physicists and astronomers to do, but at least we are no longer completely clueless about the development of the universe.

Naturally, however, we don't merely want to understand the cosmos from the time just after the Big Bang. We want to go all the way back to the beginning—and perhaps beyond. What was *before* the Big Bang?

This question, hanging somewhere between science, religion, and philosophy, occupied people long before they knew anything about the Big Bang. To begin with, there were religious creation myths, which had nothing to do with science, but at least made for a decent story. The Christian version, in which the universe is simply created by the will of God, is perhaps the least interesting of the lot. In Nordic mythology, the world was created when the first gods slew the giant Ymir and formed the earth out of his flesh, the oceans out of his blood, the mountains out of his bones, and the trees out of his hair. His skull became the sky and his brain the clouds. In Japanese Shintoism, the world was formed by the deities Izanagi and Izanami churning the chaotic primeval ocean with a heavenly, jewel-inlaid spear. The salt that then dripped from the spear became the first land, on which they built a palace in which to get married. On their wedding night, Izanagi and Izanami gave birth to the various islands of the Japanese archipelago. Meanwhile, according to the Kuba people of Central Africa, it was the giant Mbombo who created the world by vomiting up first the sun, moon, and stars and then the ancestors of all animals and humans. Then there is the ancient Egyptian cosmogony of Heliopolis, in which the god of light, Atum, simply first created himself and then produced the god and goddess Shu and Tefnut by masturbating.

When people turned from religion to science in their attempts to understand the world, they initially ignored the question of a beginning. Even until the first half of the twentieth century, the majority of people assumed that there simply was no beginning—the universe had always existed and would always do so.

Even Albert Einstein, whose relativity theory later led to the modern concept of the Big Bang, advocated this idea of a static universe. He even distrusted the equations of his own theories that showed that the universe could not be static. He modified the formulae in order to "correct" this—had he not done so, he could have predicted the later observations of Edwin Hubble (and would have become even more famous than he already was). In the 1920s, Hubble (together with his colleagues Vesto Slipher and Milton Humason) observed the very thing that Einstein was so reluctant to accept: that all galaxies are moving away from each other, and that the further apart they are, the more quickly they are moving. Put another way, the universe is expanding. And if it is expanding and will therefore be bigger in the future than in the present, it stands to reason that it must have been smaller in the past than it is today.

These findings and the analysis of Einstein's equations inspired the Belgian physicist (and theologian) Georges Lemâitre to postulate a hot and dense state as the universe's beginning. In the beginning, he suggested, all matter was concentrated in a primeval atom or "Cosmic Egg," from which today's universe burst out and then proceeded to expand until today. To the ears of most physicists, such a concept sounded suspiciously like the Christian creation myth, but the observations and, above all, Einstein's theory of relativity were unequivocal. The universe is indeed not static and is instead expanding. Even Einstein withdrew his "corrections" (which he supposedly referred to later as the "greatest blunder of my life"). In the decades that followed, there were ever more accurate theories and observations, and by 1964, when Arno Penzias and Robert Wilson discovered the cosmic background radiation predicted by the Big Bang theory, the majority of scientists were convinced of its validity. Yet as far as the ultimate beginning was concerned, just as much (or little) was known as at the time of the mythological creation stories, though this changed nothing with regard to the fundamental importance of finding the answer to the question "What was before the Big Bang?" and people's undiminished interest in this.

The only thing is, as Hawking demonstrated in his first important work, there is a singularity blocking the way between us and a possible answer to the question. At the moment of the Big Bang itself, all existing theories collapse. Whatever may have happened back then, it simply cannot be understood using the general theory of relativity.

As already mentioned, Hawking hoped to find a solution to this by taking quantum mechanics into account. If he could somehow manage to combine relativity theory with quantum mechanics, a new overarching theory could perhaps involve quantum effects that got rid of the singularity. After all, the universe gets smaller and smaller as we go back closer to the Big Bang. There must be some point where it is so tiny that it must be described as a quantum object and is therefore subject to the laws of quantum mechanics. If that were the case, then the singularity would perhaps not be a singularity after all, but rather "smeared," as it were, on account of the quantum blurring. What is valid for "particles" in quantum mechanics is also valid for the potential singularity. Just as a particle has no exact location and can instead be viewed as a wave, so too would the singularity no longer be a single point with infinite density, but rather something else—the only thing being that nobody really knows what that "something else" could be, since there is not yet a suitable "quantum-relativity theory."

Stephen Hawking took yet another route in order to attempt to solve this problem. Together with the American physicist James Hartle, he returned to an idea also originating from quantum mechanics, though he then took it further. To give a clear description of the two men's concept is, however, even more difficult than in the case of the radiation emitted by black holes, since we here have to contend with "imaginary time."

"Imaginary time"—it sounds like something that somebody simply dreamed up. The word "imaginary," however, doesn't mean that this time is something unreal or invented. It actually refers to the "imaginary numbers" which have been around in mathematics for a long time. These are numbers that fundamentally function like all other numbers, with the exception that they don't have a direct correspondence to our daily lives. We can easily imagine "four," because we simply need to think of four objects—four apples in a basket, for instance. If each apple costs a dollar, but we only have three dollars and so owe the seller a dollar after buying them, we can also quite easily imagine a negative number like –1. With imaginary numbers, on the other hand, this doesn't work so well. They are defined via the imaginary unit i, which you get from the root of the number –1. The number i is therefore a number that gives you –1 when multiplied by itself.

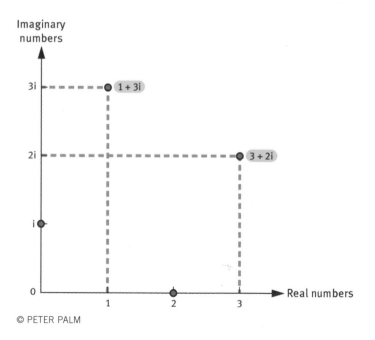

Anybody who has never come across such numbers is likely to be rather perplexed by the concept. We learn at school that every number, whether positive or negative, always produces a positive number when multiplied by itself, but nothing prevents us in principle from defining a number like i. And we can then calculate with it and create terms like $4 \times i$, or $5 + i$ or 7 divided by 3i, for instance. There are also completely normal and easily understandable calculation rules with imaginary numbers, and they fit seamlessly into the structure of the "normal" numbers. It is, however, impossible to imagine a clear representation of them, however, and we simply have to accept their existence.

We can multiply any number by the imaginary unit, and the result will be an imaginary number. If the number we are working with is one that describes time, then the result of this calculation will be an imaginary time. That may sound strange, but it is a completely normal process in natural sciences and mathematics. There are many fields in natural sciences in which we calculate using imaginary numbers, and they have turned out to be highly practical in explaining a whole load of real phenomena.

Imaginary numbers and imaginary time are needed above all to carry out many calculations in quantum mechanics, with some of these indeed possible only if imaginary time is used. Hawking and Hartle decided to investigate what imaginary time would do to the singularity of the Big Bang. As discussed in the first chapter, the universe has been described using four-dimensional space-time since Albert Einstein, made up of three space dimensions and one time dimension. If we use imaginary time instead of normal time—as Hawking and Hartle did—space-time becomes what we call "Euclidean space-time."

Simply put, multiplying time by the imaginary unit causes time to cease behaving like time and be more like a direction in the spatial dimension. Just as we can go backward and forward in the spatial dimension, we could go backward and forward in imaginary time. By using imaginary numbers, the difference between space and time disappears, and this has consequences for the singularity.

In Euclidean space-time, there is no past in which everything collapses into a single point. The singularity has disappeared. Instead, the imaginary space-time describes something that can be compared to the surface of the earth, on which even the North and South Poles are just completely normal points. As described by Hawking and Hartle, the universe functions in the exact same way. Like the surface of the earth, it is not infinitely large but does not have any boundaries. The cosmos has no "origin" in time. There is simply some point before which there is no time. And, even if we can scarcely imagine this, there is also no sensible question about "before." Just as it makes no sense to ask what lies north of the North Pole, so too is it impossible to pose a sensible question about what came before Euclidean space-time.

The universe is therefore at rest within itself. It is completely determined by itself; there is no "outside" or "before"; nothing that was in existence "before" the Big Bang happened is needed in order to get everything started, since there is no "before" the Big Bang. Space and time form a uniform surface that is boundless and yet has no boundaries (hence why Hawking and Hartle called their idea the "no-boundary proposal").

According to Hawking, the universe simply *was*, without time as we know it. It existed in "imaginary time," where there is no singularity. But if we (mathematically) transform this cosmos back into real time, then there *was* a beginning of time—the point that we call the Big Bang and which, with our current theories, looks like a singularity. This singularity was, however, not a special point—just as the earth's North Pole is simply a point, and not a singularity where peculiar things happen. Or, to put it more mathematically, it was possible using the new method to calculate the state of the universe at a time shortly after its formation, without having to worry about what happened before, simply by using this irrational time instead of normal time.

Hawking and Hartle thought up a timeless universe that nevertheless had a beginning. How wonderful! If we look far enough back into the past and come ever closer to the time of the Big Bang, then there is a point at which time—simply put—disappears and becomes space. In this imaginary space-time, the universe has no boundaries, and there is no time. The beginning of our real time was 13.8 billion years ago, and this was also the beginning of what we have observed since then: the beginning of the expanding universe, which is becoming bigger and bigger and behaves just as we observe and understand it.

All that being said, we mustn't allow ourselves to get too carried away with such demonstrations. It is only possible to truly understand the work of Hawking and Hartle on a mathematical level, not concretely. The idea of imaginary time is above all a "trick calculation" (which shouldn't sound, and is not meant to be, disparaging) and not the explanation of a concrete physical process. The universe didn't *really* run on imaginary time in the past. Hawking and Hartle developed a mathematical method that would remove the problem of the singularity caused by the general theory of relativity. It is, however, only a proposal, one that can be definitively tested only with great difficulty. What we can read into the state of the real universe using the no-boundary proposal ties in with our observations. The proposal is therefore not yet refuted, at least, which is certainly a good sign. But this does not automatically mean that it does indeed accurately describe the universe.

Nobody knows why things are supposed to have happened as Hawking and Hartle described. As Hawking himself said, this idea cannot be derived from any other physical principles. It does not necessarily follow from other already confirmed statements about the cosmos. It is a suggestion that leaves many questions unanswered. Why should there first be only space, or space with imaginary time, and then suddenly real time? What would cause the universe to behave in such a way?

Stephen Hawking was unable to answer these questions during his lifetime. But he did show us that we need not be completely clueless when it comes to the truly big questions. The question of the beginning of creation is so huge and all-encompassing that we almost treat it with too much reverence and do not dare to go in search of answers. Hawking demonstrated that we need not limit our thinking. We can and indeed should ponder questions that were traditionally left to theology. We may well not find any definitive answers (something religion also hasn't managed to do), but science is at least in a position to use its methods to set out in search of such answers. And if we should ever find them, then it would be, as Stephen Hawking says at the end of *A Brief History of Time*, "the ultimate triumph of human reason—for then we should know the mind of God."

Epilogue

THIS BOOK PROVIDES AN OVERVIEW OF STEPHEN HAWK-ing's work. During the course of his long career, however, he did of course work on many more subjects than I have presented here. A single, slim volume like this cannot do justice to such a prolific and unique scientific life. In any case, it would have to remain unfinished. When Hawking died on March 14, 2018 (Albert Einstein's 139th birthday), he had answered many questions but had left at least as many again to future generations. Physicists around the world still seek to unite quantum mechanics and relativity theory, in the hope—just like Hawking—of finally finding out what lies behind the phenomenon that is hidden by the singularities even today. Science is no less fascinated by the question of the very beginning (and what was before it) than Stephen Hawking was.

Hawking was without doubt a genius. Neverthe-less, it would be an exaggeration to rank him alongside scientists like Isaac Newton or Albert Einstein. He discovered completely new and astonishing things about our universe, but unlike his two predecessors, he didn't completely revolutionize the natural sciences. In another way, however, Hawking was indeed unique—like no other scientist, past or present, he was able to communicate the profoundly mathematical and abstract subjects of his research to the wider public.

He didn't just write a great number of books that could be understood by the average man in the street and became bestsellers—he also took theoretical physics deep into the world of popular culture. With his daughter, he created a successful series of children's books about science. He played himself in the TV series *Star Trek: The Next Generation* (something nobody else had done before), had numerous guest appearances in *The Simpsons* and *Futurama*, appeared in *The Big Bang Theory* and a Monty Python sketch, and spoke on Pink Floyd's album *The Division Bell*. Documentaries were made about him, as were TV shows and a movie, with Eddie Redmayne winning a Golden Globe and an Oscar for his portrayal of Hawking. It is no exaggeration to say that Stephen Hawking was the most famous scientist of the late twentieth and early twenty-first centuries.

Hawking recognized how important it is that the public is not completely disconnected from science. The best research is of little value if it is not communicated, and this is particularly true of the abstract basic research carried out by Stephen Hawking. His work on black holes, Euclidean space-time, or the Big Bang has no immediate effects on our everyday lives. It will not make our smartphones faster or give our computers more storage space, and we cannot cure any diseases with it. Hawking devoted his entire scientific life to the exact opposite of practical, applied research. But that was precisely what made his work so fascinating for so many people. Knowledge about black holes or the beginning of the universe is no less "practical" than a Beethoven concerto or a Leonardo da Vinci painting. The abstract world of quantum fields, singularities, and event horizons may well be incomprehensible to all but a handful of scientists, yet it nevertheless touches something that exerts a boundless fascination deep within us. For it is precisely these abstract themes that promise answers to the big questions. It is the mathematical universes that give us clues to a fundamental order underpinning the real cosmos.

What makes us human is our boundless thirst for knowledge. We want to understand the world—no, we *must* understand it. Whether or not there really is a "theory of everything"—a theory that unites quantum mechanics and relativity theory and which allows us to understand the universe in its entirety—we won't be able to give up searching for it.

When faced with such a massive undertaking, it would be easy to lose heart, particularly since the task has become significantly greater over the last few decades. Now, theoretical physics no longer has to struggle with just *one* universe. All the signs point to the existence of a "multiverse," of which our observable cosmos is merely a small part. But precisely this should actually fill us with pride, says Hawking at the end of the new edition of *A Brief History of Time*: "Despite the vastness of the multiverse, there is a sense in which we remain significant: we can still be proud to be part of a species that is working all this out."

And it is indeed highly impressive that we humans are in a position to develop such lines of thought; that there is enough room in our heads for the universe (or multiverse) in all its unimaginable magnitude and complexity and that we are able to investigate and—at least to an extent—comprehend it.

Whether the vision of Stephen Hawking and all those—past, present, and future—who tread the same path in search of a fundamental explanation of the universe will ever come to fruition is uncertain. But if we ever come up with a "theory of everything," then it should also be a "theory for everybody." This is exactly what Hawking says to lead up to the closing paragraph of *A Brief History of Time* quoted here in the last chapter, which we should therefore cite in its entirety and not merely reduce to the provocative statement about "the mind of God."

Hawking talks of the major specialization that the natural sciences have undergone in the past decades and centuries. Today, no single individual can easily maintain an overview of all research findings. It is even more difficult to explain this research in a manner that can be understood by the general public. Science advances far too quickly to allow time for popular science to come up with an explanation that can penetrate deep into the minds of readers of magazines or books—and hardly anybody has the time or inclination to process in a suitable manner everything that astrophysics, for instance, discovers on a daily basis.

The whole paragraph in Hawking's book reads: "However, if we discover a complete theory, it should in time be understandable by everyone, not just by a few scientists. Then we shall all, philosophers, scientists and just ordinary people, be able to take part in the discussion of the question of why it is that we and the universe exist. If we find the answer to that, it would be the ultimate triumph of human reason—for then we should know the mind of God."

We have yet to discover this complete theory, but in view of the significance of our search, we are actually in a win-win situation. For even if we fail, we will still make triumphant discoveries and be able to see the cosmos with completely new eyes.

Recommended
Reading in a Nutshell

BOOKS BY STEPHEN HAWKING
A Brief History of Time, London 1988

The book that made Stephen Hawking famous is as fascinating to read now as it was at the time of publication. Should you wish to treat yourself to this pleasure today, however, you should go for one of the extended new editions, in which Hawking takes into account the numerous discoveries that have been made in cosmology since 1988.

Black Holes and Baby Universes and Other Essays, New York 1993

This book contains twelve essays written between 1976 and 1992. Some of them come from lectures by Stephen Hawking that require a certain amount of background scientific knowledge to be truly understandable. Some of them deal with Hawking's private life and his illness.

The Nature of Space and Time, Princeton, New York 1996

Stephen Hawking cowrote this book with the physicist Roger Penrose. It retraces a discussion between the two men about physics and the philosophy of physics that focuses on the fundamental nature of space and time. The book is difficult to understand without prior scientific knowledge.

The Universe in a Nutshell, London 2001

The Universe in a Nutshell is Stephen Hawking's second popular science book and is also about cosmology. This time, however, the focus is less on black holes. Hawking concentrates more on the nature of time, as well as elucidating string theory and its potential ability to unite quantum mechanics and relativity theory, and addressing the issue of time travel.

A Briefer History of Time, New York 2005

Written with the physicist Leonard Mlodinow, the book is essentially a sequel to *A Brief History of Time*. More than fifteen years after Hawking's first bestseller, it brings the stories about the Big Bang, black holes, and string theory up to date and presents them in a new light.

On the Shoulders of Giants: The Great Works of Physics and Astronomy, Philadelphia 2002

In this book, Hawking presents short biographies of the great minds that came before him—Nicolaus Copernicus, Galileo Galilei, Johannes Kepler, Isaac Newton, and Albert Einstein.

The Grand Design, London 2010

Together with Leonard Mlodinow, Hawking describes the search for a "Theory of Everything"—a theory that would unite quantum mechanics and relativity theory. After an easy-to-understand discussion of the fundamental aspects of these two theories, the authors propose how the universe and its formation could be explained without having to resort to religious ideas. Hawking presents here in more detail the universe without boundaries first mentioned in *A Brief History of Time*.

My Brief History, New York 2013

Here, Hawking recounts his childhood, youth, and time as a student at Cambridge, along with his diagnosis with ALS and how he coped with the illness. The book is indeed brief but offers fascinating insights into Hawking's life.

Brief Answers to the Big Questions, New York 2018

In this posthumously published book, Hawking explains the universe for one last time. It brings together some of his texts that look at humanity's most fascinating questions, from "Is there a God?" or "Why are we here?" to "Should we colonize space?"

CHILDREN'S BOOKS BY STEPHEN HAWKING

Together with his daughter Lucy, Stephen Hawking published a series of children's books. They tell of young George and Annie, two children who live next door to each other, Annie's father and his super-computer, "Cosmos," which uncovers the universe's mysteries for the children. Cosmos helps George and Annie learn all about the solar system and the universe, quantum mechanics and relativity theory, the Big Bang and black holes, and all sorts of other phenomena and theories from the field of natural sciences. But they also have to defend the super-computer against Graham Reeper, who wants to use Cosmos for his own dark schemes.

The five volumes in the series contain challenging scientific ideas which can nevertheless be understood by children, thanks to the exciting storylines and excellent illustrations:

Volume 1: *George's Secret Key to the Universe*, New York 2007

Volume 2: *George's Cosmic Treasure Hunt*, New York 2009

Volume 3: *George and the Big Bang*, New York 2011

Volume 4: *George and the Unbreakable Code*, New York 2015

Volume 5: *George and the Blue Moon*, New York 2017

BOOK ABOUT STEPHEN HAWKING

A fascinating insight into Hawking's life seen from a different perspective is offered by *Travelling to Infinity: My Life with Stephen* (Richmond 2007) by Jane Hawking, his first wife, to whom he was married for thirty years. The book tells the story of Hawking's life and formed the basis for the 2014 film *The Theory of Everything*, which won an Oscar and two Golden Globes.